The Renewable Energy Blueprint

1

Designing a Cleaner, Greener Future

Sadie Lipsey

The Renewable Energy Blueprint

© Copyright 2024 by **Sadie Lipsey**

All rights reserved

This document is geared towards providing exact and reliable information with regards to the topic and issue covered. The publication is sold with the idea that the publisher is not required to render accounting, officially permitted, or otherwise, qualified services. If advice is necessary, legal or professional, a practiced individual in the profession should be ordered.

From a Declaration of Principles which was accepted and approved equally by a Committee of the American Bar Association and a Committee of Publishers and Associations.

In no way is it legal to reproduce, duplicate, or transmit any part of this document in either electronic means or in printed format. Recording of this publication is strictly prohibited and any storage of this document is not allowed unless with written permission from the publisher. All rights reserved.

The information provided herein is stated to be truthful and consistent, in that any liability, in terms of inattention or otherwise, by any usage or abuse of any policies, processes, or directions contained within is the solitary and utter responsibility of the recipient reader. Under no circumstances will any legal responsibility or blame be held against the publisher for any reparation, damages, or monetary loss due to the information herein, either directly or indirectly.

Respective authors own all copyrights not held by the publisher.

The information herein is offered for informational purposes solely, and is universal as so. The presentation of the information is without contract or any type of guarantee assurance.

The trademarks that are used are without any consent, and the publication of the trademark is without permission or backing by the trademark owner. All trademarks and brands within this book are for clarifying purposes only and are owned by the owners themselves, not affiliated with this document.

TABLE OF CONTENTS

Chapter 1: Solar Energy: Illuminating the Path Forward

Fundamentals of Solar Power Generation

Solar power generation stands as a beacon of hope in the quest for sustainable energy solutions. At its core, solar energy harnesses the sun's radiant light and heat, converting it into usable electricity through various technologies. The most prevalent method involves photovoltaic (PV) cells, which are semiconductor devices that directly convert sunlight into electricity. These cells are typically made from silicon, a material known for its excellent conductive properties. When sunlight strikes the PV cells, it excites electrons, creating an electric current that can be captured and used to power homes, businesses, and even entire communities.

The process begins with the absorption of photons, the basic units of light, by the PV cells. This absorption generates electron-hole pairs, which are then separated by an electric field within the cell. The movement of these electrons towards the cell's surface creates a flow of electric current. This direct current (DC) is then converted into alternating current (AC) using an inverter, making it compatible with the electrical grid and household appliances. The efficiency of this conversion process is a critical factor in the overall performance of solar panels, with advancements in technology continually pushing the boundaries of what is possible.

Beyond photovoltaic systems, solar power generation also encompasses concentrated solar power (CSP) technologies. CSP systems use mirrors or lenses to concentrate a large area of sunlight onto a small area, typically a receiver. The concentrated light is converted into heat, which drives a heat engine, often a steam turbine, connected to an electrical power generator. This method is particularly effective in regions with high direct sunlight and is often used in large-scale solar power plants.

The integration of solar power into the energy mix offers numerous benefits. It is a clean, renewable source of energy that significantly reduces greenhouse gas emissions compared to fossil fuels. Solar power systems require minimal maintenance and have a long lifespan, often exceeding 25 years. Additionally, the decentralization of energy production through solar installations can enhance energy security and resilience, reducing reliance on centralized power grids and fossil fuel imports.

However, the adoption of solar power is not without its challenges. One of the primary obstacles is the intermittent nature of solar energy, as it is dependent on weather conditions and daylight hours. This variability necessitates the development of efficient energy storage solutions, such as batteries, to ensure a stable and reliable power supply. Advances in battery technology, including lithium-ion and emerging solid-state batteries, are crucial in addressing this challenge and enabling the widespread adoption of solar energy.

Another significant barrier is the initial cost of solar installations. While the cost of solar panels has decreased dramatically over the past decade, the upfront investment can still be prohibitive for some individuals and businesses. Financial incentives, such as tax credits, rebates, and feed-in tariffs, play a vital role in making solar power more accessible and affordable. Additionally, innovative financing models, such as solar leasing and power purchase agreements (PPAs), allow consumers to adopt solar energy without the burden of high initial costs.

The environmental impact of solar power generation is generally positive, but it is essential to consider the entire lifecycle of solar technologies. The production of PV cells involves the use of hazardous materials and energy-intensive processes. Therefore, sustainable manufacturing practices and recycling programs are critical in minimizing the environmental footprint of solar technologies. Research into alternative materials, such as perovskites, offers promising avenues for reducing the environmental impact and improving the efficiency of solar cells.

The global landscape of solar power generation is diverse, with countries around the world adopting solar technologies at varying rates. Leading the charge are nations like China, the United States, and Germany, which have invested heavily in solar infrastructure and research. These countries have implemented supportive policies and incentives to drive the growth of solar energy, setting ambitious targets for renewable energy adoption. In emerging markets, solar power presents an opportunity to leapfrog traditional energy infrastructure, providing electricity to remote and underserved communities.

Technological innovations continue to shape the future of solar power generation. Bifacial solar panels, which capture sunlight on both sides, offer increased efficiency and energy yield. Floating solar farms, installed on bodies of water, maximize land use and reduce water evaporation. Building-integrated photovoltaics (BIPV) seamlessly incorporate solar cells into building materials, transforming windows, roofs, and facades into energy-generating surfaces. These advancements, coupled with ongoing research and development, hold the potential to revolutionize the solar industry and accelerate the transition to a sustainable energy future.

The role of solar power in mitigating climate change cannot be overstated. As the world grapples with the urgent need to reduce carbon emissions, solar energy offers a viable and scalable solution. By displacing fossil fuel-based electricity generation, solar power contributes to a cleaner, healthier environment. It also fosters economic growth by creating jobs in manufacturing, installation, and maintenance, driving innovation and investment in the renewable energy sector.

In conclusion, the fundamentals of solar power generation encompass a complex interplay of technology, economics, and environmental considerations. As advancements continue to unfold, the potential for solar energy to transform the global energy landscape becomes increasingly apparent. Embracing solar power is not merely a technological choice but a commitment to a sustainable and resilient future.

Technological Innovations in Solar Energy

Solar energy has undergone a remarkable transformation over the past few decades, driven by technological innovations that have significantly enhanced its efficiency, accessibility, and affordability. These advancements have not only expanded the potential applications of solar power but have also played a crucial role in accelerating the global shift towards renewable energy sources. At the heart of this transformation lies the continuous evolution of photovoltaic (PV) technology, which has seen substantial improvements in both performance and cost-effectiveness.

One of the most significant breakthroughs in solar technology is the development of high-efficiency solar cells. Traditional silicon-based solar cells have been the industry standard for years, but recent innovations have led to the creation of new materials and cell designs that offer superior performance. Among these, perovskite solar cells have garnered considerable attention due to their remarkable efficiency and low production costs. These cells are made from a class of materials known as perovskites, which have a unique crystal structure that allows for efficient light absorption and charge transport. Researchers have achieved impressive efficiency rates with perovskite cells, rivaling those of conventional silicon cells, and ongoing research aims to further enhance their stability and scalability.

Another notable advancement is the emergence of bifacial solar panels, which are capable of capturing sunlight on both sides of the panel. This design allows for increased energy generation, as the panels can harness reflected sunlight from the ground or surrounding surfaces. Bifacial panels are particularly effective in environments with high albedo, such as snowy or sandy regions, where the reflective properties of the ground can significantly boost energy output. This innovation not only maximizes the efficiency of solar installations but also reduces the overall cost of solar energy by increasing the energy yield per panel.

The integration of solar technology into everyday structures has also seen significant progress, with building-integrated photovoltaics (BIPV) leading the charge. BIPV systems incorporate solar cells into building materials, such as windows, roofs, and facades, seamlessly blending energy generation with architectural design. This approach not only enhances the aesthetic appeal of solar installations but also optimizes the use of available space, making it an attractive option for urban environments where land is limited. BIPV technology is continually evolving, with advancements in transparent solar cells and flexible materials opening up new possibilities for innovative architectural designs.

Floating solar farms represent another exciting development in the solar energy landscape. These installations consist of solar panels mounted on floating platforms, typically deployed on bodies of water such as reservoirs, lakes, or even the ocean. Floating solar farms offer several advantages, including reduced land use, decreased water evaporation, and improved panel efficiency due to the cooling effect of the water. This technology is particularly beneficial in regions with limited land availability or high population density, where traditional ground-mounted solar farms may not be feasible.

Energy storage solutions have also seen significant advancements, addressing one of the primary challenges of solar energy: its intermittent nature. The development of efficient and cost-effective battery technologies, such as lithium-ion and emerging solid-state batteries, has been instrumental in enabling the widespread adoption of solar power. These storage systems allow for the capture and retention of excess solar energy generated during peak sunlight hours, ensuring a stable and reliable power supply even when the sun is not shining. Innovations in battery technology continue to evolve, with research focused on enhancing energy density, reducing costs, and improving the lifespan of storage systems.

The digitalization of solar energy systems has further revolutionized the industry, with smart technologies playing a pivotal role in optimizing performance and efficiency. Advanced monitoring and control systems, powered by sophisticated algorithms, enable real-time data analysis and predictive maintenance, ensuring that solar installations operate at peak efficiency. These systems can detect and diagnose issues before they become critical, minimizing downtime and reducing maintenance costs. Additionally, smart inverters and grid integration technologies facilitate seamless communication between solar installations and the electrical grid, enhancing grid stability and enabling the integration of distributed energy resources.

The rise of solar energy has also been supported by innovative financing models that have made solar installations more accessible to a broader audience. Solar leasing and power purchase agreements (PPAs) allow consumers to adopt solar energy without the burden of high upfront costs, providing a flexible and affordable pathway to renewable energy adoption. These models have been instrumental in driving the growth of residential and commercial solar installations, democratizing access to clean energy and fostering a more sustainable energy future.

As the solar industry continues to evolve, research and development efforts remain focused on pushing the boundaries of what is possible. Emerging technologies, such as tandem solar cells, which combine multiple layers of different materials to capture a broader spectrum of sunlight, hold the potential to achieve unprecedented efficiency levels. Additionally, advancements in solar thermal technologies, which convert sunlight into heat for industrial processes or electricity generation, offer promising avenues for expanding the applications of solar energy beyond traditional electricity generation.

The global transition to renewable energy is an ongoing journey, and technological innovations in solar energy are at the forefront of this transformation. By continually improving the efficiency, affordability, and versatility of solar technologies, the industry is paving the way for a sustainable energy future. These advancements not only contribute to reducing carbon emissions and combating climate change but also drive economic growth by creating jobs and fostering innovation in the renewable energy sector. As we look to the future, the continued evolution of solar technology promises to play a pivotal role in shaping a cleaner, more resilient energy landscape for generations to come.

Case Studies: Successful Solar Projects Around the World

Across the globe, solar energy projects have emerged as shining examples of innovation and sustainability, demonstrating the transformative potential of harnessing the sun's power. These projects, varying in scale and scope, offer valuable insights into the successful implementation of solar technologies and the diverse benefits they bring to communities and economies.

In the heart of the Mojave Desert, the Ivanpah Solar Electric Generating System stands as a testament to the capabilities of concentrated solar power (CSP). Spanning over 3,500 acres, Ivanpah is one of the largest solar thermal power plants in the world. It utilizes thousands of mirrors, known as heliostats, to focus sunlight onto central towers, where the concentrated heat generates steam to drive turbines. This innovative approach allows Ivanpah to produce a significant amount of electricity while minimizing its environmental footprint. The project has not only contributed to California's renewable energy goals but has also provided valuable lessons in large-scale solar deployment and the integration of CSP technology into the energy mix.

In a different corner of the world, the Kamuthi Solar Power Project in Tamil Nadu, India, showcases the potential of photovoltaic technology on a massive scale. As one of the largest solar farms globally, Kamuthi covers an area of 2,500 acres and boasts a capacity of 648 megawatts. The project was completed in a record time of just eight months, highlighting the efficiency and scalability of solar PV installations. Kamuthi has played a crucial role in India's ambitious renewable energy targets, providing clean electricity to thousands of households and reducing the country's reliance on fossil fuels. The project's success underscores the importance of supportive government policies and investment in infrastructure to drive the growth of solar energy.

In the Netherlands, the SolaRoad project offers a unique perspective on integrating solar technology into everyday infrastructure. This innovative initiative involves the installation of solar panels within road surfaces, transforming them into energy-generating pathways. The pilot project, located in the town of Krommenie, has demonstrated the feasibility of this concept, generating electricity to power streetlights and other local amenities. SolaRoad exemplifies the creative application of solar technology, showcasing how existing infrastructure can be leveraged to produce renewable energy without occupying additional land. The project's success has sparked interest in similar initiatives worldwide, paving the way for further exploration of solar-integrated infrastructure.

In Africa, the Noor Ouarzazate Solar Complex in Morocco stands as a beacon of sustainable development and energy independence. Situated on the edge of the Sahara Desert, Noor is one of the largest solar complexes in the world, combining CSP and PV technologies to achieve a total capacity of over 580 megawatts. The project is a cornerstone of Morocco's renewable energy strategy, aiming to generate 52% of its electricity from renewable sources by 2030. Noor has not only reduced the country's carbon emissions but has also created jobs and stimulated economic growth in the region. The complex serves as a model for other countries seeking to harness their solar potential and transition to a sustainable energy future.

In Australia, the Hornsdale Power Reserve, often referred to as the "Tesla Big Battery," represents a groundbreaking approach to energy storage and grid stability. While not a solar project per se, the Hornsdale facility is closely linked to solar energy, providing a critical solution to the intermittency of renewable sources. Located in South Australia, the battery system stores excess solar and wind energy, releasing it during periods of high demand or low generation. This innovative project has demonstrated the effectiveness of large-scale battery storage in enhancing grid reliability and supporting the integration of renewable energy. The success of Hornsdale has inspired similar projects worldwide, highlighting the importance of energy storage in the transition to a renewable energy future.

In Chile, the Atacama Desert, known for its abundant sunlight, is home to the Cerro Dominador Solar Thermal Plant. This project combines CSP and PV technologies to harness the desert's solar potential, generating clean electricity for the national grid. Cerro Dominador features a solar tower surrounded by heliostats, similar to the Ivanpah system, and a PV array, maximizing energy production. The project is a key component of Chile's renewable energy strategy, contributing to the country's goal of achieving carbon neutrality by 2050. Cerro Dominador exemplifies the successful integration of multiple solar technologies, offering a blueprint for other regions with high solar irradiance.

In the United Arab Emirates, the Mohammed bin Rashid Al Maktoum Solar Park stands as a symbol of the nation's commitment to renewable energy. Located in Dubai, the solar park is one of the largest in the world, with a planned capacity of 5,000 megawatts by 2030. The project incorporates both PV and CSP technologies, providing a diverse and reliable energy supply. The solar park is a cornerstone of Dubai's Clean Energy Strategy, aiming to generate 75% of the city's energy from renewable sources by 2050. The project's success has positioned the UAE as a leader in solar energy, demonstrating the potential for large-scale solar deployment in arid regions.

These case studies illustrate the diverse applications and benefits of solar energy projects worldwide. From large-scale solar farms to innovative infrastructure integration, these initiatives highlight the potential of solar technology to drive sustainable development and energy transformation. Each project offers valuable lessons in overcoming challenges, leveraging local resources, and fostering collaboration between governments, businesses, and communities. As the world continues to embrace renewable energy, these successful solar projects serve as beacons of inspiration, guiding the way towards a cleaner, more sustainable future.

Overcoming Barriers to Solar Adoption

Solar energy, with its promise of clean and renewable power, has captured the imagination of many. Yet, despite its potential, several barriers have hindered its widespread adoption. Understanding and addressing these challenges is crucial for accelerating the transition to a sustainable energy future.

One of the most significant barriers to solar adoption is the initial cost of installation. While the price of solar panels has decreased significantly over the past decade, the upfront investment required for purchasing and installing a solar system can still be prohibitive for many individuals and businesses. This financial hurdle is particularly pronounced in regions with lower income levels or limited access to financing options. To overcome this barrier, innovative financing models have emerged, such as solar leasing and power purchase agreements (PPAs). These models allow consumers to adopt solar energy without the burden of high initial costs, offering flexible payment plans and long-term savings on electricity bills. Additionally, government incentives, such as tax credits, rebates, and grants, play a vital role in reducing the financial burden and making solar energy more accessible to a broader audience.

Another challenge is the intermittent nature of solar energy, which depends on weather conditions and daylight hours. This variability can lead to fluctuations in energy supply, posing a challenge for grid stability and reliability. To address this issue, advancements in energy storage technologies have become essential. Batteries, such as lithium-ion and emerging solid-state options, provide a solution by storing excess solar energy generated during peak sunlight hours for use during periods of low generation. These storage systems ensure a stable and reliable power supply, enabling the integration of solar energy into the grid. Continued research and development in battery technology are crucial for improving energy density, reducing costs, and extending the lifespan of storage solutions.

The lack of awareness and understanding of solar technology also poses a barrier to adoption. Many potential users are unfamiliar with the benefits and capabilities of solar energy, leading to misconceptions and hesitancy. Public education and outreach programs are essential for raising awareness and providing accurate information about solar technology. These initiatives can include workshops, seminars, and informational campaigns that highlight the environmental and economic benefits of solar energy. By fostering a better understanding of solar technology, these programs can empower individuals and communities to make informed decisions about adopting renewable energy solutions.

Regulatory and policy frameworks can also impact the adoption of solar energy. In some regions, complex permitting processes, restrictive zoning laws, and lack of supportive policies can hinder the development of solar projects. Streamlining regulatory processes and implementing favorable policies are critical for facilitating solar adoption. Governments can play a pivotal role by establishing clear guidelines, offering incentives, and setting ambitious renewable energy targets. Collaborative efforts between policymakers, industry stakeholders, and communities can create an enabling environment for solar energy development, driving growth and innovation in the sector.

Grid infrastructure and connectivity present another challenge, particularly in remote or underserved areas. In regions with limited grid access, the deployment of solar energy systems can be constrained by inadequate infrastructure. Off-grid and microgrid solutions offer a viable alternative, providing electricity to communities that are not connected to the central grid. These systems can be tailored to local needs, offering a decentralized approach to energy generation and distribution. By investing in grid infrastructure and supporting off-grid solutions, governments and organizations can expand access to solar energy and promote energy equity.

Environmental and social considerations also play a role in solar adoption. While solar energy is a clean and renewable source, the production and disposal of solar panels involve environmental impacts. The manufacturing process requires energy and materials, and the disposal of panels at the end of their lifecycle can pose challenges. Sustainable manufacturing practices and recycling programs are essential for minimizing the environmental footprint of solar technologies. Additionally, engaging with local communities and addressing social concerns, such as land use and cultural impacts, are crucial for ensuring the successful implementation of solar projects.

Technological advancements continue to drive the evolution of solar energy, offering solutions to overcome existing barriers. Innovations in solar cell materials, such as perovskites, promise higher efficiency and lower production costs. Bifacial panels, which capture sunlight on both sides, and building-integrated photovoltaics (BIPV), which incorporate solar cells into building materials, offer new possibilities for maximizing energy generation. These advancements, coupled with ongoing research and development, hold the potential to revolutionize the solar industry and accelerate the transition to a sustainable energy future.

Collaboration and partnerships are key to overcoming barriers to solar adoption. By working together, governments, businesses, research institutions, and communities can pool resources, share knowledge, and drive innovation. Collaborative efforts can lead to the development of new technologies, the establishment of supportive policies, and the creation of financing mechanisms that make solar energy more accessible and affordable. By fostering a culture of collaboration, stakeholders can address challenges collectively and pave the way for a cleaner, more sustainable energy landscape.

The journey towards widespread solar adoption is a complex and multifaceted endeavor, requiring concerted efforts from all sectors of society. By addressing financial, technological, regulatory, and social barriers, we can unlock the full potential of solar energy and create a brighter, more sustainable future for generations to come. As we continue to innovate and collaborate, the promise of solar energy as a cornerstone of the global energy transition becomes increasingly attainable.

The Future of Solar Energy: Trends and Predictions

The future of solar energy is poised to be a dynamic and transformative force in the global energy landscape. As the world grapples with the pressing need to transition to sustainable energy sources, solar power stands out as a beacon of hope, offering a clean, abundant, and increasingly cost-effective solution. The trends and predictions shaping the future of solar energy are driven by technological advancements, policy shifts, and evolving market dynamics, all of which promise to redefine how we generate and consume electricity.

One of the most significant trends in the solar industry is the continuous improvement in photovoltaic (PV) technology. The efficiency of solar panels has been steadily increasing, thanks to innovations in materials and cell design. Perovskite solar cells, for instance, have emerged as a promising alternative to traditional silicon-based cells, offering high efficiency and low production costs. Researchers are also exploring tandem solar cells, which combine multiple layers of different materials to capture a broader spectrum of sunlight, potentially achieving unprecedented efficiency levels. These advancements are expected to drive down the cost of solar energy further, making it more competitive with conventional energy sources.

The integration of solar energy with energy storage solutions is another critical trend shaping the future of the industry. As solar power generation is inherently intermittent, the ability to store excess energy for use during periods of low sunlight is crucial for ensuring a stable and reliable power supply. Advances in battery technology, particularly in lithium-ion and emerging solid-state batteries, are paving the way for more efficient and cost-effective storage solutions. These developments are expected to enhance the viability of solar energy as a primary power source, enabling greater penetration into the energy mix and reducing reliance on fossil fuels.

The rise of decentralized energy systems is also set to play a pivotal role in the future of solar energy. Distributed generation, where electricity is produced at or near the point of use, is gaining traction as a viable alternative to centralized power plants. This trend is driven by the increasing affordability of solar panels and the growing demand for energy independence and resilience. Residential and commercial solar installations, often paired with battery storage, allow consumers to generate their own electricity, reducing their reliance on the grid and providing a buffer against power outages. This shift towards decentralized energy systems is expected to empower individuals and communities, fostering a more resilient and sustainable energy future.

Policy and regulatory frameworks will continue to be instrumental in shaping the future of solar energy. Governments around the world are setting ambitious renewable energy targets and implementing supportive policies to drive the growth of solar power. These measures include tax incentives, feed-in tariffs, and renewable portfolio standards, all of which create a favorable environment for solar investment and development. As the urgency of addressing climate change intensifies, it is anticipated that policymakers will increasingly prioritize renewable energy, further accelerating the adoption of solar technologies.

The global expansion of solar energy is also expected to be fueled by the growing demand for clean energy in emerging markets. In regions with abundant sunlight and limited access to electricity, solar power offers a cost-effective and scalable solution for meeting energy needs. Countries in Africa, Asia, and Latin America are investing in solar infrastructure to provide electricity to underserved communities, stimulate economic growth, and reduce their carbon footprint. This trend is likely to drive significant growth in the solar industry, as emerging markets become key players in the global energy transition.

Technological innovations in solar energy are not limited to improvements in efficiency and storage. The development of smart grid technologies and digitalization is set to revolutionize the way solar energy is managed and distributed. Advanced monitoring and control systems, powered by sophisticated algorithms, enable real-time data analysis and predictive maintenance, ensuring that solar installations operate at peak efficiency. These systems facilitate seamless communication between solar installations and the electrical grid, enhancing grid stability and enabling the integration of distributed energy resources. The digitalization of solar energy systems is expected to optimize performance, reduce costs, and enhance the overall reliability of solar power.

The environmental and social benefits of solar energy are also expected to drive its future growth. As a clean and renewable source of energy, solar power significantly reduces greenhouse gas emissions, contributing to global efforts to combat climate change. The deployment of solar energy creates jobs in manufacturing, installation, and maintenance, driving economic growth and fostering innovation in the renewable energy sector. Additionally, solar energy offers a pathway to energy equity, providing electricity to remote and underserved communities and reducing energy poverty.

Looking ahead, the future of solar energy is bright, with the potential to transform the global energy landscape and drive the transition to a sustainable energy future. The trends and predictions shaping the industry are underpinned by a commitment to innovation, collaboration, and sustainability. As technological advancements continue to unfold, the cost of solar energy is expected to decrease further, making it an increasingly attractive option for consumers and businesses alike. The integration of solar energy with storage solutions and smart grid technologies will enhance its viability as a primary power source, enabling greater penetration into the energy mix and reducing reliance on fossil fuels.

The global expansion of solar energy, driven by demand in emerging markets and supportive policy frameworks, will further accelerate the adoption of solar technologies. As the world embraces the potential of solar power, the industry is poised to play a pivotal role in shaping a cleaner, more resilient energy future for generations to come. The journey towards a sustainable energy future is a complex and multifaceted endeavor, requiring concerted efforts from all sectors of society. By addressing challenges, leveraging local resources, and fostering collaboration, we can unlock the full potential of solar energy and create a brighter, more sustainable future.

Chapter 2: Wind Energy: Riding the Currents of Change

The Science Behind Wind Power

Wind power, a cornerstone of renewable energy, harnesses the kinetic energy of moving air to generate electricity. This process, rooted in the principles of physics and engineering, has evolved significantly over the years, transforming wind into a reliable and efficient source of power. Understanding the science behind wind power involves exploring the mechanics of wind turbines, the factors influencing wind energy production, and the technological advancements that have enhanced its viability.

At the heart of wind power generation are wind turbines, which convert the kinetic energy of wind into mechanical energy and subsequently into electrical energy. A typical wind turbine consists of several key components: the rotor blades, the nacelle, and the tower. The rotor blades, usually three in number, are designed to capture the wind's energy. As wind flows over the blades, it creates a pressure difference, causing the blades to rotate. This rotation drives a shaft connected to a generator housed within the nacelle, converting mechanical energy into electricity. The tower elevates the rotor and nacelle to capture stronger and more consistent winds found at higher altitudes.

The efficiency of a wind turbine is influenced by several factors, including wind speed, air density, and blade design. Wind speed is the most critical factor, as the power generated by a turbine is proportional to the cube of the wind speed. This means that even small increases in wind speed can lead to significant increases in energy output. Consequently, wind turbines are often sited in locations with high average wind speeds, such as coastal areas, open plains, and hilltops. Air density, which varies with altitude and temperature, also affects energy production. Higher air density results in greater energy capture, making cooler and lower-altitude locations more favorable for wind power generation.

Blade design plays a crucial role in the efficiency of wind turbines. Modern blades are typically made from lightweight materials such as fiberglass or carbon fiber, allowing them to capture wind energy effectively while minimizing structural stress. The shape and angle of the blades, known as the airfoil and pitch, are optimized to maximize lift and minimize drag, enhancing the turbine's ability to convert wind energy into rotational motion. Advances in computational fluid dynamics and materials science have led to the development of more efficient blade designs, contributing to the overall performance of wind turbines.

The science of wind power also involves understanding the variability and intermittency of wind energy. Unlike fossil fuels, wind energy is not constant, as it depends on weather patterns and geographical factors. This variability poses challenges for grid integration and energy reliability. To address these challenges, wind power systems are often combined with energy storage solutions, such as batteries or pumped hydro storage, which store excess energy generated during periods of high wind for use during low-wind periods. Additionally, advancements in forecasting technology have improved the ability to predict wind patterns, enabling better planning and management of wind energy resources.

Technological innovations have played a pivotal role in advancing wind power, making it a more competitive and sustainable energy source. One such innovation is the development of offshore wind farms, which take advantage of the stronger and more consistent winds found over open water. Offshore wind turbines are typically larger and more powerful than their onshore counterparts, capable of generating significant amounts of electricity. The deployment of floating wind turbines, which can be anchored in deeper waters, has further expanded the potential for offshore wind energy, opening up new areas for development.

The integration of digital technologies and smart grid systems has also enhanced the efficiency and reliability of wind power. Advanced monitoring and control systems enable real-time data analysis and predictive maintenance, ensuring that wind turbines operate at peak performance. These systems can detect and diagnose issues before they become critical, minimizing downtime and reducing maintenance costs. Smart grid technologies facilitate seamless communication between wind farms and the electrical grid, enhancing grid stability and enabling the integration of distributed energy resources.

The environmental benefits of wind power are significant, as it produces no greenhouse gas emissions during operation and has a relatively small land footprint compared to other energy sources. Wind farms can coexist with agricultural and recreational land uses, providing a sustainable and versatile energy solution. However, the environmental impact of wind power is not negligible, as the production and disposal of turbine components involve energy and materials. Sustainable manufacturing practices and recycling programs are essential for minimizing the environmental footprint of wind power technologies.

The social and economic benefits of wind power are also noteworthy. The deployment of wind energy creates jobs in manufacturing, installation, and maintenance, driving economic growth and fostering innovation in the renewable energy sector. Wind power offers a pathway to energy independence, reducing reliance on imported fossil fuels and enhancing energy security. Additionally, wind energy provides a source of income for landowners and communities hosting wind farms, contributing to local economic development.

The future of wind power is bright, with continued advancements in technology and policy support driving its growth. The development of larger and more efficient turbines, coupled with innovations in energy storage and grid integration, promises to enhance the viability of wind energy as a primary power source. As the world seeks to transition to a sustainable energy future, wind power is poised to play a crucial role in reducing carbon emissions and combating climate change.

The science behind wind power is a testament to human ingenuity and the potential of renewable energy to transform the global energy landscape. By harnessing the power of the wind, we can create a cleaner, more sustainable future for generations to come. The journey towards widespread adoption of wind energy is a complex and multifaceted endeavor, requiring concerted efforts from all sectors of society. By addressing challenges, leveraging local resources, and fostering collaboration, we can unlock the full potential of wind power and create a brighter, more sustainable future.

Onshore vs. Offshore Wind Farms: A Comparative Analysis

Wind energy has emerged as a pivotal player in the global shift towards renewable energy, with both onshore and offshore wind farms contributing significantly to electricity generation. Each type of wind farm offers unique advantages and challenges, shaped by their distinct environments and technological requirements. A comparative analysis of onshore and offshore wind farms reveals the nuances of their development, operation, and impact, providing valuable insights for stakeholders considering investment in wind energy.

Onshore wind farms, located on land, are the more traditional form of wind energy generation. They have been instrumental in the early adoption of wind power, benefiting from relatively lower installation and maintenance costs compared to their offshore counterparts. The accessibility of onshore sites simplifies logistics, allowing for easier transportation of materials and equipment. This accessibility also facilitates routine maintenance and repairs, contributing to lower operational costs. Onshore wind farms can be integrated into existing land uses, such as agriculture, allowing for dual-purpose land utilization and providing additional income streams for landowners.

However, onshore wind farms face several challenges, primarily related to site selection and community acceptance. Suitable sites for onshore wind farms are often limited by geographical and environmental constraints, such as topography, land availability, and proximity to transmission infrastructure. Additionally, onshore wind farms can encounter resistance from local communities due to concerns about noise, visual impact, and potential effects on wildlife. Addressing these concerns requires careful planning, community engagement, and adherence to environmental regulations to ensure that onshore wind projects are developed sustainably and with minimal disruption.

Offshore wind farms, situated in bodies of water, offer a different set of advantages and challenges. One of the most significant benefits of offshore wind farms is the availability of stronger and more consistent wind resources compared to onshore sites. This results in higher capacity factors and greater energy output, making offshore wind an attractive option for large-scale power generation. The vast expanses of open water also allow for the deployment of larger turbines, which can capture more energy and contribute to economies of scale.

The challenges associated with offshore wind farms are primarily related to the complexity and cost of installation and maintenance. Offshore environments present harsh conditions, requiring specialized equipment and techniques for construction and operation. The logistics of transporting materials and personnel to offshore sites can be complex and costly, and maintenance activities are often constrained by weather conditions and accessibility. Additionally, the development of offshore wind farms requires significant investment in grid infrastructure to transmit electricity from remote offshore locations to onshore demand centers.

Environmental considerations play a crucial role in the development of both onshore and offshore wind farms. Onshore wind farms must navigate potential impacts on local ecosystems, including effects on bird and bat populations, habitat disruption, and land use changes. Mitigation measures, such as careful site selection, turbine design modifications, and habitat restoration, are essential for minimizing these impacts. Offshore wind farms, while generally having a lower impact on terrestrial ecosystems, must consider the effects on marine life, including fish, marine mammals, and seabirds. The construction and operation of offshore wind farms can alter marine habitats and affect species behavior, necessitating comprehensive environmental assessments and monitoring programs.

The economic implications of onshore and offshore wind farms are also a key consideration. Onshore wind farms typically have lower capital costs and shorter development timelines, making them an attractive option for investors seeking quicker returns. The established nature of onshore wind technology and the availability of mature supply chains further contribute to cost-effectiveness. Offshore wind farms, while requiring higher initial investment, offer the potential for significant energy generation and long-term economic benefits. The development of offshore wind can stimulate economic growth through job creation in manufacturing, construction, and maintenance, as well as through the development of supporting industries and infrastructure.

The choice between onshore and offshore wind farms is influenced by a range of factors, including geographical location, available resources, regulatory frameworks, and market conditions. In regions with limited land availability or high population density, offshore wind may offer a viable alternative to meet renewable energy targets. Conversely, in areas with abundant land resources and favorable wind conditions, onshore wind farms may provide a more cost-effective solution. The decision-making process involves a careful evaluation of the trade-offs between cost, energy output, environmental impact, and community acceptance.

Technological advancements continue to shape the future of both onshore and offshore wind energy. Innovations in turbine design, materials, and installation techniques are driving improvements in efficiency and cost-effectiveness. For onshore wind farms, advancements in turbine technology, such as taller towers and longer blades, are enabling access to higher wind speeds and increased energy capture. In the offshore sector, the development of floating wind turbines is expanding the potential for wind energy in deeper waters, where fixed-bottom structures are not feasible. These technological innovations are expected to enhance the competitiveness of wind energy and support its continued growth as a key component of the global energy transition.

The integration of wind energy into the broader energy system is also a critical consideration. Both onshore and offshore wind farms contribute to grid stability and energy security, providing a renewable and domestic source of electricity. The variability of wind energy requires effective grid management and the integration of complementary technologies, such as energy storage and demand response, to ensure a reliable power supply. The development of smart grid systems and digital technologies is facilitating the efficient integration of wind energy, enabling real-time monitoring and control of energy flows.

The comparative analysis of onshore and offshore wind farms highlights the diverse opportunities and challenges associated with each type of wind energy generation. Both onshore and offshore wind farms play a vital role in the transition to a sustainable energy future, offering complementary solutions to meet the growing demand for clean electricity. By understanding the unique characteristics and considerations of each type of wind farm, stakeholders can make informed decisions that maximize the benefits of wind energy while minimizing its impacts. As the world continues to embrace renewable energy, the synergy between onshore and offshore wind farms will be instrumental in achieving a resilient and sustainable energy system.

Environmental and Social Impacts of Wind Energy

Wind energy, a cornerstone of the renewable energy revolution, offers a sustainable alternative to fossil fuels, yet it is not without its environmental and social implications. Understanding these impacts is crucial for developing wind energy projects that are both effective and responsible. By examining the environmental and social dimensions of wind energy, stakeholders can make informed decisions that balance the benefits of clean energy with the need to protect ecosystems and communities.

The environmental impacts of wind energy are multifaceted, encompassing both positive and negative aspects. On the positive side, wind energy significantly reduces greenhouse gas emissions, contributing to global efforts to combat climate change. Unlike fossil fuel-based power generation, wind turbines produce electricity without releasing carbon dioxide or other harmful pollutants, making them a clean and sustainable energy source. Additionally, wind energy has a relatively small land footprint compared to other energy sources, allowing for dual land use in agricultural or grazing areas.

However, the development and operation of wind farms can also have adverse environmental effects. One of the most prominent concerns is the impact on wildlife, particularly birds and bats. Wind turbines can pose a collision risk for flying animals, leading to fatalities and potential disruptions to local ecosystems. To mitigate these impacts, developers can implement measures such as careful site selection, turbine design modifications, and operational adjustments, such as curtailing turbine operation during peak migration periods. Ongoing research and monitoring are essential for understanding and minimizing the effects of wind energy on wildlife.

The construction of wind farms can also lead to habitat disruption and land use changes. The installation of turbines, access roads, and transmission lines can alter landscapes and affect local flora and fauna. To address these concerns, developers can conduct thorough environmental assessments and engage in habitat restoration efforts to minimize the ecological footprint of wind projects. By prioritizing biodiversity conservation and sustainable land management practices, wind energy can coexist with natural ecosystems.

The social impacts of wind energy are equally important, influencing public perception and acceptance of wind projects. Wind farms can bring economic benefits to local communities, including job creation, increased tax revenue, and opportunities for local businesses. The construction and maintenance of wind farms require a skilled workforce, providing employment opportunities in rural and remote areas. Additionally, landowners hosting wind turbines can receive lease payments, offering a stable source of income and contributing to local economic development.

Despite these benefits, wind energy projects can also face social challenges, particularly related to community acceptance and engagement. Concerns about noise, visual impact, and potential health effects can lead to opposition from local residents. Effective communication and community involvement are crucial for addressing these concerns and building trust with stakeholders. Developers can engage with communities early in the planning process, providing transparent information and opportunities for public participation. By fostering open dialogue and collaboration, wind energy projects can gain social license and support from local communities.

The visual impact of wind turbines is another consideration, as they can alter the aesthetic character of landscapes. While some people view wind turbines as symbols of progress and sustainability, others may perceive them as intrusive or unsightly. Balancing the need for renewable energy with the preservation of cultural and scenic values requires careful planning and design. Developers can work with local communities and stakeholders to identify suitable locations and design solutions that minimize visual impact while maximizing energy generation.

The integration of wind energy into the broader energy system also has social implications, particularly related to energy equity and access. Wind energy can contribute to energy independence and security, reducing reliance on imported fossil fuels and enhancing resilience to energy price fluctuations. By providing a domestic and renewable source of electricity, wind energy can help diversify the energy mix and support the transition to a low-carbon economy. However, ensuring that the benefits of wind energy are equitably distributed requires attention to issues of access and affordability, particularly for marginalized and underserved communities.

The development of wind energy projects can also intersect with cultural and historical considerations, particularly in regions with indigenous or traditional land use practices. Respecting cultural heritage and engaging with indigenous communities are essential for ensuring that wind projects are developed in a socially responsible manner. Developers can collaborate with indigenous groups to incorporate traditional knowledge and values into project planning and decision-making, fostering mutual respect and understanding.

The environmental and social impacts of wind energy are complex and interconnected, requiring a holistic approach to project development and management. By addressing these impacts through careful planning, stakeholder engagement, and adaptive management, wind energy can be harnessed in a way that maximizes its benefits while minimizing its drawbacks. The transition to renewable energy is a critical component of global efforts to combat climate change and achieve sustainable development, and wind energy has a vital role to play in this transition.

As the wind energy sector continues to grow, ongoing research and innovation will be essential for enhancing the sustainability and social acceptance of wind projects. Advances in turbine technology, environmental monitoring, and community engagement practices can help address the challenges associated with wind energy development. By fostering collaboration between industry, government, and communities, stakeholders can work together to create a future where wind energy contributes to a cleaner, more equitable, and sustainable world.

The journey towards a sustainable energy future is a shared responsibility, requiring concerted efforts from all sectors of society. By understanding and addressing the environmental and social impacts of wind energy, we can unlock its full potential and create a brighter, more sustainable future for generations to come. The balance between harnessing the power of the wind and preserving the integrity of our natural and social environments is a delicate one, but it is a balance that can be achieved through thoughtful and responsible action.

Policy Frameworks Supporting Wind Energy Development

Policy frameworks play a crucial role in shaping the development and expansion of wind energy, providing the necessary support and incentives to drive investment and innovation. These frameworks encompass a range of legislative, regulatory, and financial mechanisms designed to promote the adoption of wind power and facilitate its integration into the energy mix. Understanding the various policy instruments and their impact on wind energy development is essential for stakeholders seeking to navigate the complex landscape of renewable energy policy.

One of the most effective policy tools for supporting wind energy development is the implementation of renewable energy targets. These targets, often set at national or regional levels, establish clear goals for the proportion of energy to be generated from renewable sources, including wind. By setting ambitious targets, governments signal their commitment to renewable energy, providing a stable and predictable environment for investors and developers. Renewable energy targets can drive the deployment of wind energy projects by creating demand for clean electricity and encouraging the development of supporting infrastructure.

Feed-in tariffs (FITs) have been instrumental in promoting wind energy by guaranteeing fixed payments to renewable energy producers for the electricity they generate. These payments, typically offered over a long-term contract, provide financial certainty and reduce investment risk, making wind projects more attractive to investors. FITs have been successful in accelerating the deployment of wind energy in many countries, particularly in the early stages of market development. However, as the cost of wind energy has decreased, some regions have transitioned to alternative support mechanisms, such as auctions or competitive bidding processes, to ensure cost-effectiveness and market efficiency.

Renewable portfolio standards (RPS) or renewable energy mandates are another key policy instrument used to support wind energy development. RPS require utilities to obtain a certain percentage of their electricity from renewable sources, creating a market for wind energy and other renewables. By mandating the inclusion of renewable energy in the energy mix, RPS drive demand for wind power and incentivize utilities to invest in wind projects. These standards can be tailored to regional conditions, allowing for flexibility in meeting renewable energy goals while ensuring a diverse and resilient energy supply.

Tax incentives and subsidies are commonly used to reduce the financial burden of wind energy projects and encourage investment. These incentives can take various forms, including production tax credits (PTCs), investment tax credits (ITCs), and accelerated depreciation. PTCs provide financial benefits based on the amount of electricity generated by a wind project, while ITCs offer a credit based on the initial capital investment. Accelerated depreciation allows developers to recover the cost of their investment more quickly, improving project economics. By reducing the overall cost of wind energy, tax incentives and subsidies make wind projects more competitive with conventional energy sources.

Grid access and infrastructure development are critical components of policy frameworks supporting wind energy. Ensuring that wind projects have access to the electrical grid is essential for delivering electricity to consumers and maximizing the benefits of wind power. Policies that facilitate grid connection and expansion, such as streamlined permitting processes and investment in transmission infrastructure, are vital for integrating wind energy into the broader energy system. Additionally, policies that promote the development of smart grid technologies and energy storage solutions can enhance the reliability and efficiency of wind energy, enabling greater penetration into the energy mix.

Research and development (R&D) funding is another important aspect of policy frameworks supporting wind energy. By investing in R&D, governments can drive technological innovation and improve the performance and cost-effectiveness of wind energy technologies. R&D funding can support a wide range of activities, from basic research on wind dynamics and turbine design to the development of advanced materials and control systems. By fostering innovation, R&D funding helps to overcome technical challenges and unlock new opportunities for wind energy deployment.

International cooperation and collaboration are also key elements of policy frameworks supporting wind energy. By working together, countries can share best practices, harmonize standards, and leverage collective expertise to accelerate the deployment of wind energy. International agreements and partnerships, such as those facilitated by organizations like the International Renewable Energy Agency (IRENA) and the Global Wind Energy Council (GWEC), provide platforms for knowledge exchange and capacity building. These collaborations can help to address common challenges, such as grid integration and environmental impacts, and promote the global expansion of wind energy.

Public engagement and education are essential components of policy frameworks supporting wind energy development. By raising awareness of the benefits of wind energy and addressing misconceptions, policymakers can build public support and acceptance for wind projects. Public engagement initiatives can include community consultations, educational campaigns, and stakeholder workshops, providing opportunities for dialogue and collaboration. By fostering a positive public perception of wind energy, policymakers can create a supportive environment for wind project development and ensure that the transition to renewable energy is inclusive and equitable.

The effectiveness of policy frameworks supporting wind energy development depends on their ability to adapt to changing market conditions and technological advancements. As the cost of wind energy continues to decline and new technologies emerge, policymakers must be prepared to adjust their strategies and instruments to ensure continued growth and competitiveness. This may involve transitioning from fixed incentives to market-based mechanisms, such as auctions or green certificates, to drive cost reductions and innovation. By maintaining a flexible and responsive policy environment, governments can support the long-term sustainability and success of the wind energy sector.

The development of policy frameworks supporting wind energy is a dynamic and ongoing process, requiring collaboration and coordination among a wide range of stakeholders. By aligning policy objectives with market realities and technological advancements, policymakers can create a conducive environment for wind energy development and contribute to the global transition to a sustainable energy future. The journey towards a renewable energy future is a shared responsibility, and by working together, governments, industry, and communities can harness the power of the wind to create a cleaner, more resilient, and sustainable world.

Innovations and Future Directions in Wind Technology

Wind technology has undergone remarkable advancements over the past few decades, transforming from a niche energy source into a cornerstone of the global renewable energy landscape. As the demand for clean energy continues to rise, innovations in wind technology are paving the way for even greater efficiency, reliability, and scalability. These innovations are not only enhancing the performance of wind turbines but also expanding the potential applications of wind energy, offering exciting possibilities for the future.

One of the most significant areas of innovation in wind technology is the development of larger and more efficient wind turbines. Modern turbines are designed to capture more energy from the wind by increasing the size of the rotor blades and the height of the towers. Larger rotor diameters allow turbines to sweep a greater area, capturing more wind energy and increasing electricity generation. Taller towers enable access to stronger and more consistent wind resources found at higher altitudes, further boosting energy output. These advancements in turbine design are driving down the cost of wind energy, making it more competitive with traditional energy sources.

The use of advanced materials is another key innovation in wind technology. Lightweight and durable materials, such as carbon fiber and advanced composites, are being used to construct rotor blades, reducing weight and enhancing performance. These materials allow for longer blades that can capture more wind energy without compromising structural integrity. Additionally, innovations in materials science are leading to the development of more resilient components that can withstand harsh environmental conditions, extending the lifespan of wind turbines and reducing maintenance costs.

The integration of digital technologies and data analytics is revolutionizing the operation and maintenance of wind farms. Advanced sensors and monitoring systems are being deployed to collect real-time data on turbine performance, weather conditions, and grid interactions. This data is analyzed using sophisticated algorithms to optimize turbine operation, predict maintenance needs, and enhance overall efficiency. Predictive maintenance, enabled by data analytics, allows operators to identify potential issues before they become critical, minimizing downtime and reducing operational costs. The use of digital twins—virtual replicas of physical assets—enables detailed simulations and analysis, supporting informed decision-making and continuous improvement.

Floating wind technology represents a groundbreaking innovation that is expanding the potential for offshore wind energy. Traditional offshore wind turbines are anchored to the seabed, limiting their deployment to relatively shallow waters. Floating wind turbines, however, are mounted on buoyant platforms that can be anchored in deeper waters, opening up vast new areas for wind energy development. This technology is particularly promising for regions with deep coastal waters, where fixed-bottom turbines are not feasible. Floating wind farms can take advantage of stronger and more consistent wind resources found further offshore, increasing energy generation and reducing visual impact on coastal communities.

Hybrid wind systems, which combine wind energy with other renewable energy sources, are emerging as a promising solution for enhancing grid stability and reliability. By integrating wind power with solar energy, battery storage, or other technologies, hybrid systems can provide a more consistent and reliable power supply. These systems can balance the variability of wind energy with complementary energy sources, ensuring a steady flow of electricity to the grid. Hybrid wind systems are particularly valuable in remote or off-grid locations, where they can provide a sustainable and independent energy solution.

The development of small-scale and distributed wind energy systems is another exciting direction for wind technology. These systems, designed for residential, commercial, or community use, offer a decentralized approach to energy generation. Small wind turbines can be installed on rooftops, in backyards, or in community spaces, providing clean energy directly to consumers.

Distributed wind systems can enhance energy resilience and reduce reliance on centralized power grids, offering a flexible and adaptable solution for diverse energy needs.

Innovations in wind technology are also driving improvements in grid integration and energy storage. As the penetration of wind energy increases, effective grid management and storage solutions are essential for maintaining grid stability and reliability. Advanced grid technologies, such as smart grids and demand response systems, enable better coordination between wind farms and the electrical grid, optimizing energy flows and reducing congestion. Energy storage solutions, such as batteries and pumped hydro storage, provide a means of storing excess wind energy for use during periods of low wind, ensuring a consistent power supply.

The future of wind technology is bright, with ongoing research and development efforts focused on overcoming current challenges and unlocking new opportunities. Innovations in turbine design, materials, and digital technologies are expected to continue driving down the cost of wind energy, making it an increasingly attractive option for energy generation. The expansion of offshore and floating wind technologies will open up new markets and regions for wind energy development, while hybrid and distributed systems will enhance the flexibility and resilience of the energy system.

Collaboration and partnerships between industry, academia, and government are essential for advancing wind technology and achieving a sustainable energy future. By working together, stakeholders can share knowledge, resources, and expertise, accelerating the pace of innovation and deployment. International cooperation and knowledge exchange can help to address common challenges, such as grid integration and environmental impacts, and promote the global expansion of wind energy.

The journey towards a sustainable energy future is a shared responsibility, and wind technology has a vital role to play in this transition. By harnessing the power of the wind, we can create a cleaner, more resilient, and sustainable world for generations to come. The innovations and future directions in wind technology offer exciting possibilities for the continued growth and success of wind energy, providing a pathway to a brighter and more sustainable future.

Chapter 3: Hydropower: Channeling Earth's Waterways

Principles of Hydroelectric Power Generation

Hydroelectric power generation stands as one of the oldest and most reliable forms of renewable energy, harnessing the kinetic energy of flowing water to produce electricity. This method of power generation is rooted in the fundamental principles of physics and engineering, offering a sustainable and efficient means of meeting energy demands. Understanding the principles of hydroelectric power generation provides insight into its operation, benefits, and potential challenges.

At the heart of hydroelectric power generation is the conversion of potential energy stored in water at a height into kinetic energy as it flows downward, which is then transformed into mechanical energy by a turbine and ultimately into electrical energy by a generator. This process begins with the accumulation of water in a reservoir, typically created by constructing a dam across a river. The dam serves as a barrier, elevating the water level and creating a significant difference in height, known as the hydraulic head. The potential energy of the stored water is directly proportional to the height of the water column and the volume of water available.

When water is released from the reservoir, it flows through penstocks—large pipes or tunnels—towards the turbines. The force of the moving water spins the turbine blades, converting the kinetic energy of the water into mechanical energy. The design and efficiency of the turbine are crucial, as they determine how effectively the energy of the water is harnessed. Common types of turbines used in hydroelectric power plants include Francis, Kaplan, and Pelton turbines, each suited to different head and flow conditions. Francis turbines are versatile and widely used for medium head applications, while Kaplan turbines are ideal for low head and high flow scenarios. Pelton turbines, on the other hand, are designed for high head and low flow conditions, utilizing a series of buckets to capture the energy of water jets.

The mechanical energy generated by the spinning turbine is transferred to a generator, where it is converted into electrical energy. The generator operates on the principle of electromagnetic induction, where the rotation of a coil within a magnetic field induces an electric current. This process is highly efficient, with modern hydroelectric power plants achieving conversion efficiencies of over 90%. The electricity produced is then transmitted through power lines to homes, businesses, and industries, providing a reliable and renewable source of energy.

One of the key advantages of hydroelectric power generation is its ability to provide flexible and responsive electricity supply. Hydroelectric plants can quickly adjust their output to match fluctuations in electricity demand, making them valuable assets for grid stability and reliability. This flexibility is particularly important in balancing the variability of other renewable energy sources, such as wind and solar power. Additionally, hydroelectric power plants can serve as energy storage systems through pumped storage technology. During periods of low electricity demand, excess energy can be used to pump water back into the reservoir, storing it for future use when demand is higher.

The environmental benefits of hydroelectric power generation are significant, as it produces electricity without emitting greenhouse gases or other pollutants. This makes it an attractive option for reducing carbon emissions and combating climate change. However, the construction and operation of hydroelectric power plants can have environmental and social impacts that must be carefully managed. The creation of reservoirs can lead to the flooding of large areas, affecting ecosystems, wildlife habitats, and human communities. Changes in river flow and sediment transport can also impact downstream ecosystems and water quality.

To mitigate these impacts, careful planning and environmental assessments are essential in the development of hydroelectric projects. Strategies such as fish ladders and bypass systems can help maintain aquatic biodiversity by allowing fish to migrate past dams. Additionally, adaptive management practices can be implemented to monitor and address environmental changes over time. Engaging with local communities and stakeholders is also crucial to ensure that the social and economic benefits of hydroelectric projects are equitably distributed.

The principles of hydroelectric power generation are grounded in the efficient use of natural resources, offering a sustainable and reliable means of meeting energy needs. As the world continues to transition towards renewable energy, hydroelectric power will play a vital role in providing clean and flexible electricity. By understanding the fundamental principles and addressing the associated challenges, we can harness the full potential of hydroelectric power to create a more sustainable and resilient energy future.

The Role of Small-Scale Hydropower in Local Communities

Small-scale hydropower, often referred to as micro or mini hydropower, plays a pivotal role in empowering local communities by providing a sustainable and decentralized energy solution. Unlike large-scale hydroelectric projects, which require significant infrastructure and can have substantial environmental impacts, small-scale hydropower systems are designed to harness the energy of flowing water in a way that is both environmentally friendly and socially beneficial. These systems offer a range of advantages that make them particularly well-suited for rural and remote areas, where access to reliable electricity can be limited.

The essence of small-scale hydropower lies in its ability to generate electricity from modest water flows, such as streams or small rivers, without the need for large dams or reservoirs. This approach minimizes the ecological footprint of the installation, preserving the natural landscape and aquatic ecosystems. By utilizing run-of-the-river designs, small-scale hydropower systems divert a portion of the water flow through a turbine before returning it to the river, ensuring minimal disruption to the environment. This method not only protects local biodiversity but also maintains the natural beauty of the area, which can be crucial for communities that rely on tourism or fishing.

One of the most significant benefits of small-scale hydropower is its potential to provide energy independence and security to local communities. By generating electricity locally, communities can reduce their reliance on centralized power grids and fossil fuels, enhancing their resilience to energy price fluctuations and supply disruptions. This independence is particularly valuable in regions where grid infrastructure is underdeveloped or unreliable, offering a stable and continuous power supply that can support essential services such as healthcare, education, and communication.

The economic impact of small-scale hydropower on local communities is profound. By providing a reliable source of electricity, these systems can stimulate economic development and improve living standards. Access to electricity enables the operation of small businesses, agricultural processing, and other income-generating activities, creating jobs and boosting local economies. Furthermore, the construction and maintenance of small-scale hydropower systems can provide employment opportunities for local residents, fostering skills development and capacity building within the community.

Community involvement and ownership are central to the success of small-scale hydropower projects. By engaging local stakeholders in the planning, development, and operation of these systems, communities can ensure that the projects align with their needs and priorities. This participatory approach fosters a sense of ownership and responsibility, increasing the likelihood of long-term success and sustainability. Community-owned hydropower projects can also generate revenue through the sale of excess electricity to the grid, providing a source of income that can be reinvested in local development initiatives.

The social benefits of small-scale hydropower extend beyond economic development, contributing to improved quality of life and social cohesion. Access to electricity can enhance educational opportunities by enabling the use of digital technologies and extending study hours. It can also improve healthcare services by powering medical equipment and refrigeration for vaccines and medicines. Additionally, electrification can reduce the burden of household chores, such as water pumping and food processing, freeing up time for education and leisure activities.

Despite its many advantages, the implementation of small-scale hydropower projects can face challenges that must be carefully managed. Technical and financial barriers, such as the initial cost of installation and the need for specialized expertise, can pose obstacles to project development. To overcome these challenges, communities can seek support from government programs, non-governmental organizations, and international development agencies that provide funding, technical assistance, and capacity-building resources. By leveraging these resources, communities can enhance their ability to plan, finance, and operate small-scale hydropower systems effectively.

Environmental considerations are also important in the development of small-scale hydropower projects. While these systems have a lower environmental impact than large-scale projects, careful site selection and design are essential to minimize any potential negative effects on local ecosystems. Environmental impact assessments and community consultations can help identify and address potential concerns, ensuring that projects are developed in a way that respects and preserves the natural environment.

The role of small-scale hydropower in local communities is multifaceted, offering a sustainable and empowering energy solution that can drive social and economic development. By harnessing the power of water in a way that is both environmentally and socially responsible, small-scale hydropower systems can transform the lives of individuals and communities, providing a pathway to a more sustainable and equitable future. As the world continues to seek solutions to the challenges of energy access and climate change, small-scale hydropower stands out as a promising and adaptable option that can make a meaningful difference in the lives of people around the globe.

Environmental Considerations and Mitigation Efforts

Environmental considerations are paramount in the development and operation of renewable energy projects, including wind and hydroelectric power. While these energy sources offer significant benefits in terms of reducing greenhouse gas emissions and reliance on fossil fuels, they also present unique environmental challenges that must be addressed to ensure sustainable development. Mitigation efforts play a crucial role in minimizing the ecological footprint of renewable energy projects and preserving the integrity of natural ecosystems.

One of the primary environmental concerns associated with wind energy is its impact on wildlife, particularly birds and bats. Wind turbines, with their large rotating blades, can pose a collision risk for flying animals, leading to fatalities and potential disruptions to local ecosystems. To mitigate these impacts, developers can implement a range of strategies, including careful site selection to avoid areas with high bird and bat activity, turbine design modifications to reduce collision risk, and operational adjustments such as curtailing turbine operation during peak migration periods. Ongoing research and monitoring are essential for understanding the effects of wind energy on wildlife and developing effective mitigation measures.

The construction and operation of wind farms can also lead to habitat disruption and land use changes. The installation of turbines, access roads, and transmission lines can alter landscapes and affect local flora and fauna. To address these concerns, developers can conduct thorough environmental assessments and engage in habitat restoration efforts to minimize the ecological footprint of wind projects. By prioritizing biodiversity conservation and sustainable land management practices, wind energy can coexist with natural ecosystems.

Hydroelectric power, while offering a clean and renewable energy source, presents its own set of environmental challenges. The creation of reservoirs for hydroelectric projects can lead to the flooding of large areas, affecting ecosystems, wildlife habitats, and human communities. Changes in river flow and sediment transport can also impact downstream ecosystems and water quality. To mitigate these impacts, careful planning and environmental assessments are essential in the development of hydroelectric projects. Strategies such as fish ladders and bypass systems can help maintain aquatic biodiversity by allowing fish to migrate past dams. Additionally, adaptive management practices can be implemented to monitor and address environmental changes over time.

The social and cultural impacts of renewable energy projects are also important considerations. The development of wind and hydroelectric projects can intersect with cultural and historical considerations, particularly in regions with indigenous or traditional land use practices. Respecting cultural heritage and engaging with indigenous communities are essential for ensuring that renewable energy projects are developed in a socially responsible manner. Developers can collaborate with indigenous groups to incorporate traditional knowledge and values into project planning and decision-making, fostering mutual respect and understanding.

Public engagement and education are critical components of environmental mitigation efforts. By raising awareness of the benefits and potential impacts of renewable energy projects, policymakers and developers can build public support and acceptance. Public engagement initiatives can include community consultations, educational campaigns, and stakeholder workshops, providing opportunities for dialogue and collaboration. By fostering a positive public perception of renewable energy, stakeholders can create a supportive environment for project development and ensure that the transition to renewable energy is inclusive and equitable.

The integration of renewable energy into the broader energy system also has environmental implications, particularly related to grid integration and energy storage. As the penetration of renewable energy increases, effective grid management and storage solutions are essential for maintaining grid stability and reliability. Advanced grid technologies, such as smart grids and demand response systems, enable better coordination between renewable energy sources and the electrical grid, optimizing energy flows and reducing congestion. Energy storage solutions, such as batteries and pumped hydro storage, provide a means of storing excess energy for use during periods of low generation, ensuring a consistent power supply.

The effectiveness of environmental mitigation efforts depends on their ability to adapt to changing conditions and technological advancements. As renewable energy technologies continue to evolve, stakeholders must be prepared to adjust their strategies and instruments to ensure continued growth and sustainability. This may involve transitioning from fixed incentives to market-based mechanisms, such as auctions or green certificates, to drive cost reductions and innovation. By maintaining a flexible and responsive policy environment, governments can support the long-term sustainability and success of the renewable energy sector.

Collaboration and partnerships between industry, academia, and government are essential for advancing environmental mitigation efforts and achieving a sustainable energy future. By working together, stakeholders can share knowledge, resources, and expertise, accelerating the pace of innovation and deployment. International cooperation and knowledge exchange can help to address common challenges, such as grid integration and environmental impacts, and promote the global expansion of renewable energy.

The journey towards a sustainable energy future is a shared responsibility, and environmental considerations and mitigation efforts are vital components of this transition. By understanding and addressing the environmental impacts of renewable energy projects, we can unlock their full potential and create a brighter, more sustainable future for generations to come. The balance between harnessing the power of renewable energy and preserving the integrity of our natural and social environments is a delicate one, but it is a balance that can be achieved through thoughtful and responsible action.

Advances in Marine and Hydrokinetic Energy

Marine and hydrokinetic energy represent a frontier in the realm of renewable energy, harnessing the vast and largely untapped power of the world's oceans, rivers, and tidal currents. These energy sources offer immense potential for sustainable power generation, leveraging the natural movements of water to produce electricity. As technology advances, the ability to efficiently capture and convert marine and hydrokinetic energy is becoming increasingly feasible, opening new avenues for clean energy development.

The ocean, with its relentless waves and powerful tides, holds a tremendous amount of kinetic energy. Wave energy converters are designed to capture this energy by utilizing the up-and-down motion of waves. These devices come in various forms, including point absorbers, oscillating water columns, and attenuators, each tailored to specific wave conditions and environments. Point absorbers are buoy-like structures that float on the surface, moving with the waves to drive hydraulic pumps or generators. Oscillating water columns use the rise and fall of water within a chamber to compress air, which then drives a turbine. Attenuators are long, segmented devices that lie parallel to the wave direction, flexing with the waves to generate power. These technologies are being tested and refined to maximize efficiency and durability in the harsh marine environment.

Tidal energy, another promising form of marine energy, exploits the gravitational pull of the moon and sun, which causes the rise and fall of sea levels. Tidal stream generators, akin to underwater wind turbines, are placed in areas with strong tidal currents. As water flows past the turbine blades, it generates electricity. Tidal range technologies, such as tidal barrages and lagoons, capture the potential energy created by the difference in height between high and low tides. These systems involve constructing barriers or enclosures that trap water at high tide and release it through turbines at low tide. The predictability of tides makes tidal energy a reliable and consistent source of power, offering a significant advantage over other intermittent renewable energy sources.

River and stream energy, often referred to as hydrokinetic energy, harnesses the flow of water in rivers and streams without the need for large dams or reservoirs. This approach minimizes environmental impact while providing a decentralized energy solution. Hydrokinetic turbines are submerged in the water, capturing the kinetic energy of flowing water to generate electricity. These systems are particularly well-suited for remote or off-grid locations, where they can provide a sustainable and independent energy source. The simplicity and scalability of hydrokinetic systems make them an attractive option for communities seeking to enhance energy resilience and reduce reliance on fossil fuels.

The development of marine and hydrokinetic energy technologies is not without challenges. The harsh and unpredictable marine environment poses significant technical and engineering hurdles, requiring robust and resilient designs that can withstand extreme conditions. Corrosion, biofouling, and mechanical wear are common issues that must be addressed to ensure the longevity and reliability of marine energy devices. Advances in materials science and engineering are critical to overcoming these challenges, with research focused on developing corrosion-resistant materials, innovative coatings, and advanced monitoring systems.

Environmental considerations are also paramount in the deployment of marine and hydrokinetic energy systems. The potential impacts on marine ecosystems, including changes in water flow, sediment transport, and habitat disruption, must be carefully assessed and managed. Environmental impact assessments and monitoring programs are essential for understanding and mitigating these effects, ensuring that marine energy projects are developed in a sustainable and responsible manner. Collaboration with marine biologists, ecologists, and local communities can provide valuable insights and support the development of best practices for environmental stewardship.

The integration of marine and hydrokinetic energy into the broader energy system presents both opportunities and challenges. The variability of wave and tidal energy requires effective grid management and storage solutions to ensure a stable and reliable power supply. Advanced grid technologies, such as smart grids and demand response systems, can enhance the integration of marine energy by optimizing energy flows and reducing congestion. Energy storage solutions, such as batteries and pumped hydro storage, provide a means of storing excess energy for use during periods of low generation, ensuring a consistent power supply.

The economic potential of marine and hydrokinetic energy is significant, offering opportunities for job creation, economic development, and energy security. The growth of the marine energy sector can stimulate innovation and investment, driving the development of new technologies and industries. By harnessing the power of the ocean and rivers, countries can diversify their energy portfolios, reduce dependence on imported fuels, and enhance energy resilience.

International collaboration and knowledge exchange are essential for advancing marine and hydrokinetic energy technologies. By sharing research, expertise, and best practices, countries can accelerate the development and deployment of these technologies, overcoming common challenges and unlocking new opportunities. Organizations such as the International Renewable Energy Agency (IRENA) and the Ocean Energy Systems (OES) provide platforms for cooperation and capacity building, supporting the global expansion of marine energy.

The future of marine and hydrokinetic energy is promising, with ongoing research and development efforts focused on improving efficiency, reducing costs, and enhancing environmental sustainability. As technology continues to evolve, the potential for marine energy to contribute to a sustainable and resilient energy future becomes increasingly attainable. By harnessing the power of the ocean and rivers, we can create a cleaner, more sustainable world, providing a pathway to a brighter and more resilient energy future. The journey towards unlocking the full potential of marine and hydrokinetic energy is a shared responsibility, requiring collaboration, innovation, and commitment from all stakeholders.

The Future of Hydropower in a Sustainable Energy Mix

Hydropower has long been a cornerstone of renewable energy, providing a reliable and efficient source of electricity for over a century. As the world transitions towards a more sustainable energy future, the role of hydropower is evolving, adapting to new challenges and opportunities within the energy landscape. The future of hydropower in a sustainable energy mix is shaped by technological advancements, environmental considerations, and the integration of complementary renewable energy sources.

One of the key factors influencing the future of hydropower is the development of innovative technologies that enhance efficiency and reduce environmental impact. Advances in turbine design, such as fish-friendly turbines and variable-speed generators, are improving the ecological compatibility of hydropower projects. These technologies minimize the impact on aquatic life and ecosystems, addressing one of the primary environmental concerns associated with traditional hydropower. Additionally, the use of advanced materials and manufacturing techniques is increasing the durability and performance of hydropower infrastructure, extending the lifespan of existing facilities and reducing maintenance costs.

The integration of digital technologies and data analytics is transforming the operation and management of hydropower plants. Smart monitoring systems and predictive maintenance tools enable operators to optimize performance, reduce downtime, and enhance grid stability. By leveraging real-time data and advanced algorithms, hydropower facilities can respond more effectively to fluctuations in electricity demand and integrate seamlessly with other renewable energy sources. This digital transformation is not only improving the efficiency of hydropower but also supporting the broader transition to a smart and resilient energy grid.

The role of hydropower in a sustainable energy mix is also influenced by its ability to complement and balance other renewable energy sources, such as wind and solar power. As the penetration of variable renewables increases, the need for flexible and reliable energy sources becomes more critical. Hydropower, with its ability to provide both base-load and peak-load power, is uniquely positioned to support grid stability and reliability. Pumped storage hydropower, in particular, offers a valuable solution for energy storage, allowing excess electricity to be stored and released when needed. This capability enhances the integration of intermittent renewables, ensuring a consistent and reliable power supply.

The future of hydropower is also shaped by the growing emphasis on sustainability and environmental stewardship. As awareness of the ecological and social impacts of large-scale hydropower projects increases, there is a shift towards more sustainable and community-focused approaches. Small-scale and run-of-the-river hydropower systems are gaining traction as environmentally friendly alternatives that minimize habitat disruption and preserve natural landscapes. These systems offer a decentralized energy solution that can empower local communities and support rural development.

The potential for retrofitting and upgrading existing hydropower facilities is another important aspect of the future energy mix. Many hydropower plants around the world are aging and in need of modernization to meet current efficiency and environmental standards. By investing in upgrades and retrofits, operators can enhance the performance and sustainability of existing infrastructure, extending its useful life and maximizing its contribution to the energy mix. This approach not only reduces the need for new construction but also leverages existing resources and minimizes environmental impact.

The integration of hydropower with other renewable energy sources is a promising avenue for enhancing sustainability and resilience. Hybrid systems that combine hydropower with solar, wind, or biomass energy can provide a more balanced and reliable power supply, optimizing the use of available resources. These systems can be tailored to the specific needs and conditions of each location, offering a flexible and adaptable solution for diverse energy challenges. By leveraging the complementary strengths of different renewables, hybrid systems can enhance energy security and reduce reliance on fossil fuels.

The future of hydropower is also influenced by policy and regulatory frameworks that support sustainable energy development. Governments and international organizations play a crucial role in shaping the energy landscape through incentives, standards, and regulations that promote renewable energy and environmental protection. By creating a supportive policy environment, stakeholders can encourage investment in hydropower and other renewables, driving innovation and deployment. Collaboration and partnerships between industry, academia, and government are essential for advancing hydropower technology and achieving a sustainable energy future.

Public engagement and education are critical components of the transition to a sustainable energy mix. By raising awareness of the benefits and potential impacts of hydropower, policymakers and developers can build public support and acceptance. Public engagement initiatives can include community consultations, educational campaigns, and stakeholder workshops, providing opportunities for dialogue and collaboration. By fostering a positive public perception of hydropower, stakeholders can create a supportive environment for project development and ensure that the transition to renewable energy is inclusive and equitable.

The future of hydropower in a sustainable energy mix is characterized by innovation, integration, and collaboration. By embracing new technologies, enhancing environmental sustainability, and fostering partnerships, hydropower can continue to play a vital role in the global energy transition. As the world moves towards a cleaner and more resilient energy future, hydropower offers a pathway to achieving energy security, reducing carbon emissions, and supporting sustainable development. The journey towards a sustainable energy future is a shared responsibility, and hydropower is poised to be a key contributor to this transition, providing a reliable and sustainable source of energy for generations to come.

Chapter 4: Geothermal Energy: Unlocking Earth's Thermal Potential

Basics of Geothermal Energy Systems

Geothermal energy systems harness the Earth's natural heat to generate electricity and provide heating and cooling solutions. This renewable energy source is derived from the thermal energy stored beneath the Earth's crust, which is continuously replenished by the decay of radioactive elements and the residual heat from the planet's formation. Understanding the basics of geothermal energy systems involves exploring the various technologies, applications, and benefits associated with this sustainable energy source.

At the core of geothermal energy systems is the concept of tapping into the Earth's internal heat. This heat is accessible through geothermal reservoirs, which are areas of the Earth's crust where heat is concentrated and can be extracted for use. These reservoirs are typically located in regions with high tectonic activity, such as along plate boundaries or in volcanic areas. However, geothermal energy can also be harnessed in other regions through enhanced geothermal systems (EGS), which involve artificially creating or enhancing reservoirs by injecting water into hot, dry rock formations.

Geothermal energy systems can be broadly categorized into three main types: direct use applications, geothermal heat pumps, and geothermal power plants. Each type utilizes the Earth's heat in different ways, offering a range of solutions for various energy needs.

Direct use applications involve the direct extraction and utilization of geothermal heat for purposes such as heating buildings, greenhouses, and aquaculture facilities. This approach is one of the oldest and simplest forms of geothermal energy use, dating back thousands of years. In modern applications, geothermal heat is often used in district heating systems, where hot water from geothermal wells is distributed through a network of pipes to provide heating for multiple buildings or entire communities. This method is highly efficient and cost-effective, particularly in regions with abundant geothermal resources.

Geothermal heat pumps, also known as ground-source heat pumps, are a versatile and efficient technology for heating and cooling buildings. These systems take advantage of the relatively constant temperature of the ground or groundwater to transfer heat to or from a building. In the winter, the heat pump extracts heat from the ground and transfers it indoors, while in the summer, the process is reversed, with heat being extracted from the building and transferred back into the ground. Geothermal heat pumps are highly efficient, using significantly less energy than conventional heating and cooling systems, and can be installed in a wide range of climates and locations.

Geothermal power plants generate electricity by converting the thermal energy from geothermal reservoirs into mechanical energy, which is then used to drive a generator. There are three main types of geothermal power plants: dry steam, flash steam, and binary cycle. Dry steam plants are the simplest and oldest type, using steam directly from geothermal reservoirs to turn turbines. Flash steam plants, the most common type, use high-pressure hot water from the reservoir, which is depressurized or "flashed" into steam to drive the turbines. Binary cycle plants use a secondary fluid with a lower boiling point than water, which is vaporized by the geothermal heat and used to turn the turbines. This type of plant is particularly suited for lower-temperature geothermal resources and has the advantage of being able to operate in a closed-loop system, minimizing environmental impact.

The benefits of geothermal energy systems are numerous, making them an attractive option for sustainable energy development. Geothermal energy is a reliable and consistent source of power, providing baseload electricity generation with high capacity factors. Unlike solar and wind energy, which are dependent on weather conditions, geothermal energy is available 24/7, offering a stable and predictable energy supply. Additionally, geothermal energy systems have a relatively small environmental footprint, producing low levels of greenhouse gas emissions and requiring less land than other renewable energy sources.

The economic advantages of geothermal energy are also significant. While the initial costs of developing geothermal projects can be high, the long-term operational and maintenance costs are relatively low, resulting in competitive electricity prices. Geothermal energy can also contribute to energy security and independence by reducing reliance on imported fuels and diversifying the energy mix. Furthermore, the development of geothermal resources can stimulate local economies by creating jobs and supporting infrastructure development.

Despite its many advantages, the deployment of geothermal energy systems faces several challenges that must be addressed to realize their full potential. The exploration and development of geothermal resources can be complex and costly, requiring significant investment in drilling and resource assessment. The availability of geothermal resources is also geographically limited, with the most favorable conditions found in specific regions. To overcome these challenges, continued research and development are essential to improve exploration techniques, reduce costs, and expand the range of viable geothermal resources.

Policy and regulatory frameworks play a crucial role in supporting the growth of geothermal energy systems. Governments can encourage the development of geothermal projects through incentives, subsidies, and streamlined permitting processes. International cooperation and knowledge exchange can also facilitate the expansion of geothermal energy by sharing best practices and lessons learned from successful projects.

Public awareness and acceptance are important factors in the successful deployment of geothermal energy systems. By educating communities about the benefits and potential impacts of geothermal energy, stakeholders can build public support and foster a positive perception of this renewable energy source. Community engagement and participation in project development can also enhance social acceptance and ensure that geothermal projects align with local needs and priorities.

The future of geothermal energy systems is promising, with ongoing advancements in technology and increasing recognition of their role in a sustainable energy future. By harnessing the Earth's natural heat, geothermal energy offers a pathway to reducing carbon emissions, enhancing energy security, and supporting economic development. As the world continues to seek solutions to the challenges of climate change and energy transition, geothermal energy systems stand out as a reliable and sustainable option that can contribute to a cleaner and more resilient energy future.

Advantages and Challenges of Geothermal Energy

Geothermal energy, a cornerstone of renewable energy solutions, offers a unique blend of advantages and challenges that shape its role in the global energy landscape. As the world seeks sustainable alternatives to fossil fuels, understanding these facets of geothermal energy becomes crucial for its effective deployment and integration.

One of the most compelling advantages of geothermal energy is its reliability. Unlike solar and wind energy, which are subject to weather variability, geothermal energy provides a consistent and stable power supply. This reliability stems from the Earth's constant internal heat, which is unaffected by external climatic conditions. As a result, geothermal power plants can operate continuously, providing baseload electricity that supports grid stability and reduces the need for backup power sources.

The environmental benefits of geothermal energy are significant. Geothermal power plants emit minimal greenhouse gases compared to fossil fuel-based power generation, contributing to the reduction of carbon emissions and helping mitigate climate change. Additionally, geothermal energy has a relatively small land footprint, requiring less space than solar or wind farms. This compactness makes it an attractive option for regions with limited available land or those seeking to preserve natural landscapes.

Geothermal energy also offers economic advantages, particularly in terms of long-term cost-effectiveness. While the initial investment for geothermal projects can be high, the operational and maintenance costs are relatively low. Geothermal power plants have long lifespans, often exceeding 30 years, and benefit from stable fuel costs, as the Earth's heat is a free and inexhaustible resource. This economic stability can lead to competitive electricity prices and contribute to energy security by reducing reliance on imported fuels.

The versatility of geothermal energy extends beyond electricity generation. Geothermal resources can be used for direct heating applications, such as district heating systems, greenhouse agriculture, and industrial processes. Geothermal heat pumps provide efficient heating and cooling solutions for residential and commercial buildings, leveraging the Earth's stable temperatures to reduce energy consumption and costs. This versatility enhances the appeal of geothermal energy as a comprehensive solution for diverse energy needs.

Despite its many advantages, geothermal energy faces several challenges that must be addressed to unlock its full potential. One of the primary challenges is the geographical limitation of geothermal resources. High-quality geothermal reservoirs are typically located in regions with significant tectonic activity, such as along plate boundaries or in volcanic areas. This geographical constraint limits the widespread deployment of geothermal energy and necessitates the development of enhanced geothermal systems (EGS) to access resources in less favorable locations.

The exploration and development of geothermal resources can be complex and costly. Identifying viable geothermal sites requires extensive geological surveys, drilling, and resource assessment, which can be time-consuming and expensive. The uncertainty associated with resource exploration poses financial risks for developers and can deter investment in geothermal projects. To mitigate these risks, advancements in exploration technologies and techniques are essential, along with supportive policy frameworks that incentivize investment and reduce barriers to entry.

Environmental and social considerations also play a critical role in the development of geothermal energy projects. While geothermal energy is generally considered environmentally friendly, the extraction and use of geothermal resources can have localized impacts, such as land subsidence, induced seismicity, and the release of trace gases. Comprehensive environmental impact assessments and monitoring programs are necessary to identify and mitigate these potential effects, ensuring that geothermal projects are developed sustainably and responsibly.

Community engagement and public perception are important factors in the successful deployment of geothermal energy. Building public support and acceptance requires transparent communication about the benefits and potential impacts of geothermal projects. Engaging with local communities and stakeholders throughout the project development process can foster trust and collaboration, ensuring that projects align with local needs and priorities. Public education and awareness campaigns can also enhance the understanding of geothermal energy and its role in a sustainable energy future.

The integration of geothermal energy into the broader energy system presents both opportunities and challenges. As the penetration of variable renewable energy sources, such as solar and wind, increases, the need for flexible and reliable energy sources becomes more critical. Geothermal energy, with its ability to provide baseload power, can complement these variable sources and support grid stability. However, effective grid management and infrastructure development are necessary to facilitate the integration of geothermal energy and optimize its contribution to the energy mix.

The future of geothermal energy is promising, with ongoing research and development efforts focused on overcoming existing challenges and enhancing the efficiency and sustainability of geothermal technologies. Innovations in drilling techniques, reservoir management, and power plant design are expanding the range of viable geothermal resources and reducing costs. International collaboration and knowledge exchange are essential for advancing geothermal energy and sharing best practices and lessons learned from successful projects.

Policy and regulatory frameworks play a crucial role in supporting the growth of geothermal energy. Governments can encourage the development of geothermal projects through incentives, subsidies, and streamlined permitting processes. By creating a supportive policy environment, stakeholders can drive investment in geothermal energy and accelerate its deployment. Collaboration between industry, academia, and government is essential for advancing geothermal technology and achieving a sustainable energy future.

The journey towards a sustainable energy future is a shared responsibility, and geothermal energy is poised to be a key contributor to this transition. By understanding and addressing the advantages and challenges of geothermal energy, we can unlock its full potential and create a cleaner, more resilient energy system. As the world continues to seek solutions to the challenges of climate change and energy transition, geothermal energy offers a reliable and sustainable option that can contribute to a brighter and more sustainable future for generations to come.

Global Leaders in Geothermal Development

Geothermal energy, a cornerstone of sustainable power, has seen significant advancements and adoption across the globe. Several countries have emerged as leaders in geothermal development, each contributing uniquely to the growth and innovation of this renewable energy sector. These nations have harnessed their geological advantages, invested in technology, and implemented supportive policies to become pioneers in geothermal energy.

Iceland stands as a beacon of geothermal success, leveraging its volcanic landscape to produce nearly 90% of its heating and a significant portion of its electricity from geothermal sources. The country's commitment to renewable energy is evident in its extensive use of geothermal heat for district heating systems, which provide warmth to homes, businesses, and even greenhouses. Iceland's expertise in geothermal technology has made it a global leader, offering valuable insights and technical assistance to other countries seeking to develop their geothermal resources. The nation's success is rooted in its strategic investment in research and development, as well as its collaborative approach to sharing knowledge and best practices.

The United States, with its vast and diverse geothermal resources, is another key player in the global geothermal landscape. The country boasts the largest installed geothermal capacity in the world, with significant developments in states like California, Nevada, and Utah. The U.S. has been at the forefront of geothermal innovation, pioneering technologies such as enhanced geothermal systems (EGS) and binary cycle power plants. These advancements have expanded the range of viable geothermal resources, enabling the exploitation of lower-temperature reservoirs and increasing the efficiency of power generation. The U.S. government's support through research funding, tax incentives, and regulatory frameworks has been instrumental in driving the growth of the geothermal sector.

Kenya has emerged as a leader in geothermal development within Africa, capitalizing on the geothermal potential of the East African Rift Valley. The country's commitment to renewable energy is reflected in its ambitious plans to increase geothermal capacity, which currently accounts for a significant portion of its electricity generation. Kenya's success in geothermal development is attributed to its proactive approach to resource exploration, investment in infrastructure, and collaboration with international partners. The country's flagship geothermal project, the Olkaria Geothermal Plant, is one of the largest in the world and serves as a model for other African nations seeking to harness their geothermal resources.

The Philippines, located along the Pacific Ring of Fire, is another prominent player in the geothermal sector. The country's abundant geothermal resources have made it one of the largest producers of geothermal energy globally. The Philippines has successfully integrated geothermal power into its energy mix, reducing its dependence on imported fossil fuels and enhancing energy security. The government's supportive policies, including incentives for renewable energy development and streamlined permitting processes, have facilitated the growth of the geothermal industry. The Philippines' experience in geothermal development offers valuable lessons for other countries with similar geological conditions.

New Zealand, with its rich geothermal resources, has long been a leader in geothermal energy production. The country's geothermal power plants contribute significantly to its electricity supply, providing a reliable and sustainable energy source. New Zealand's success in geothermal development is underpinned by its commitment to environmental stewardship and community engagement. The country's geothermal projects are developed in close collaboration with local communities and indigenous groups, ensuring that the benefits of geothermal energy are shared equitably. New Zealand's approach to geothermal development emphasizes sustainability, innovation, and respect for cultural heritage.

Indonesia, home to some of the world's most active volcanoes, has immense geothermal potential. The country has made significant strides in geothermal development, with ambitious plans to expand its geothermal capacity and reduce its reliance on coal-fired power. Indonesia's government has implemented policies to attract investment in the geothermal sector, including feed-in tariffs, tax incentives, and risk mitigation measures. The country's commitment to geothermal energy is driven by its desire to achieve energy security, reduce greenhouse gas emissions, and support economic growth. Indonesia's experience highlights the importance of policy support and investment in infrastructure for successful geothermal development.

Japan, despite its limited land area, has made notable progress in geothermal energy development. The country's geothermal resources are concentrated in its volcanic regions, offering significant potential for power generation and direct use applications. Japan's government has prioritized geothermal energy as part of its renewable energy strategy, providing financial support for research and development, as well as incentives for project development. The country's focus on innovation and technology has led to advancements in geothermal exploration techniques and power plant design. Japan's experience demonstrates the potential for geothermal energy to contribute to a diversified and resilient energy mix.

Turkey, situated on the geologically active Anatolian Plate, has rapidly expanded its geothermal capacity in recent years. The country's geothermal resources are primarily used for electricity generation and direct heating applications, contributing to its renewable energy goals. Turkey's government has implemented policies to support geothermal development, including feed-in tariffs and investment incentives. The country's success in geothermal energy is driven by its strategic focus on resource exploration, infrastructure development, and international collaboration. Turkey's experience underscores the importance of a supportive policy environment and investment in research and development for geothermal growth.

These global leaders in geothermal development offer valuable insights and lessons for other countries seeking to harness their geothermal resources. Their experiences highlight the importance of strategic investment, supportive policies, and international collaboration in driving the growth of the geothermal sector. By sharing knowledge and best practices, these nations can help accelerate the global transition to renewable energy and contribute to a more sustainable and resilient energy future. As the world continues to seek solutions to the challenges of climate change and energy transition, geothermal energy stands out as a reliable and sustainable option that can play a vital role in achieving a cleaner and more sustainable future for all.

The journey of geothermal energy development is not without its hurdles, but the experiences of these leading nations provide a roadmap for overcoming challenges and maximizing the potential of this renewable resource. One of the critical factors in successful geothermal development is the commitment to research and innovation. By investing in cutting-edge technologies and exploration techniques, countries can unlock new geothermal resources and improve the efficiency of existing systems. This commitment to innovation is evident in the advancements made by the United States and Japan, where enhanced geothermal systems and advanced power plant designs have expanded the range of viable geothermal projects.

Another essential element is the establishment of a supportive policy and regulatory framework. Governments play a crucial role in creating an environment conducive to geothermal development by offering incentives, streamlining permitting processes, and providing financial support for research and infrastructure. The experiences of countries like the Philippines and Turkey demonstrate the impact of favorable policies in attracting investment and accelerating project development. By fostering a stable and predictable policy environment, governments can encourage private sector participation and drive the growth of the geothermal industry.

Technological Innovations in Geothermal Exploration

Geothermal exploration has long been a field of intrigue and potential, offering a sustainable energy source that taps into the Earth's natural heat. As the demand for renewable energy grows, technological innovations in geothermal exploration are paving the way for more efficient and effective utilization of this resource. These advancements are transforming the way geothermal energy is discovered, assessed, and harnessed, making it a more viable option for a broader range of applications and locations.

One of the most significant technological innovations in geothermal exploration is the development of advanced geophysical survey techniques. These methods allow for a more precise understanding of subsurface conditions, enabling explorers to identify geothermal reservoirs with greater accuracy. Techniques such as seismic reflection and refraction, magnetotellurics, and gravity surveys provide detailed images of the Earth's crust, revealing the presence of heat sources and fluid pathways. By integrating these geophysical methods with geological and geochemical data, researchers can create comprehensive models of geothermal systems, reducing the uncertainty and risk associated with exploration.

The use of remote sensing technology has also revolutionized geothermal exploration. Satellite imagery and aerial surveys offer a bird's-eye view of potential geothermal sites, allowing for the identification of surface features indicative of geothermal activity, such as hot springs, fumaroles, and altered vegetation. These remote sensing tools can cover large areas quickly and cost-effectively, providing valuable information for preliminary site assessments. Additionally, thermal infrared imaging can detect surface temperature anomalies, offering clues about underlying geothermal activity.

Drilling technology has seen remarkable advancements, significantly impacting the efficiency and cost-effectiveness of geothermal exploration. Innovations such as directional drilling and coiled tubing drilling have improved the ability to reach and assess geothermal reservoirs. Directional drilling allows for precise targeting of geothermal resources, enabling the exploration of complex geological formations and reducing the environmental footprint of drilling operations. Coiled tubing drilling offers a faster and more flexible alternative to traditional drilling methods, reducing the time and cost associated with well construction.

Enhanced geothermal systems (EGS) represent a groundbreaking innovation in the field, expanding the potential for geothermal energy beyond naturally occurring reservoirs. EGS involves creating or enhancing geothermal reservoirs by injecting water into hot, dry rock formations, stimulating the flow of heat to the surface. This technology has the potential to unlock vast geothermal resources in regions previously considered unsuitable for development. By overcoming the limitations of conventional geothermal systems, EGS can significantly increase the availability and accessibility of geothermal energy.

The integration of data analytics and machine learning is transforming the way geothermal exploration is conducted. These technologies enable the analysis of large datasets from various sources, identifying patterns and correlations that may not be apparent through traditional methods. Machine learning algorithms can predict the location and characteristics of geothermal reservoirs, optimizing exploration efforts and reducing the risk of unsuccessful drilling. By leveraging the power of data, geothermal explorers can make more informed decisions and improve the overall success rate of their projects.

Microseismic monitoring is another innovative technique that has gained traction in geothermal exploration. This method involves the detection and analysis of small-scale seismic events, providing insights into subsurface processes and reservoir dynamics. Microseismic monitoring can help identify areas of increased permeability and fluid flow, offering valuable information for reservoir management and development. This technology also plays a crucial role in ensuring the safety and sustainability of geothermal operations by monitoring induced seismicity and assessing potential risks.

The development of advanced materials and drilling fluids has further enhanced the efficiency and sustainability of geothermal exploration. High-temperature-resistant materials and drilling fluids are essential for operating in the extreme conditions encountered in geothermal wells. Innovations in these materials have improved the durability and performance of drilling equipment, reducing maintenance costs and extending the lifespan of geothermal infrastructure. Environmentally friendly drilling fluids have also been developed to minimize the ecological impact of drilling operations, aligning with the broader goals of sustainable energy development.

Collaboration and knowledge sharing among industry, academia, and government are driving innovation in geothermal exploration. Research institutions and universities play a vital role in advancing geothermal technology, conducting studies and experiments that push the boundaries of what is possible. Industry partnerships and government support provide the resources and infrastructure needed to bring these innovations to market. By fostering a collaborative environment, stakeholders can accelerate the development and deployment of new technologies, ensuring that geothermal energy remains a competitive and sustainable option in the global energy mix.

The future of geothermal exploration is bright, with ongoing advancements poised to unlock new opportunities and applications. As technology continues to evolve, the potential for geothermal energy to contribute to a sustainable energy future becomes increasingly apparent. By embracing innovation and leveraging the latest tools and techniques, the geothermal industry can overcome existing challenges and expand its reach, providing a reliable and renewable energy source for generations to come. The journey of geothermal exploration is one of discovery and progress, driven by the relentless pursuit of knowledge and the desire to harness the Earth's natural power for the benefit of all.

As geothermal exploration advances, the integration of cutting-edge technologies promises to redefine the boundaries of what is achievable. One area of burgeoning interest is the application of artificial intelligence in predictive modeling. By simulating geothermal systems and their responses to various stimuli, AI can offer insights into reservoir behavior, optimizing extraction processes and enhancing efficiency. These models can predict how a reservoir will respond to different extraction techniques, allowing for more strategic planning and resource management.

Another promising development is the use of fiber optic technology for real-time monitoring of geothermal wells.

Distributed temperature sensing (DTS) and distributed acoustic sensing (DAS) systems utilize fiber optic cables to provide continuous data on temperature and acoustic signals within a well. This real-time monitoring capability allows for immediate detection of changes in reservoir conditions, enabling operators to make timely adjustments and maintain optimal performance. The ability to monitor wells continuously also enhances safety by providing early warnings of potential issues such as wellbore instability or equipment failure.

The exploration of deep geothermal resources is gaining momentum, driven by the potential to access higher temperatures and greater energy yields. Advances in drilling technology, such as high-temperature drilling systems and improved drill bit materials, are making it feasible to reach depths previously considered unattainable. Deep geothermal exploration holds the promise of tapping into vast energy reserves, offering a sustainable solution for regions with limited surface geothermal resources. As these technologies mature, they could revolutionize the geothermal industry by opening up new frontiers for energy production.

Integrating Geothermal Energy into National Grids

Integrating geothermal energy into national grids presents a unique opportunity to enhance energy security, reduce carbon emissions, and provide a stable power supply. As countries strive to diversify their energy portfolios and transition to renewable sources, geothermal energy offers a reliable and sustainable solution. The process of integrating geothermal energy into national grids involves several key considerations, including grid infrastructure, regulatory frameworks, and technological advancements.

One of the primary advantages of geothermal energy is its ability to provide baseload power. Unlike intermittent renewable sources such as solar and wind, geothermal power plants can operate continuously, delivering a consistent and stable supply of electricity. This characteristic makes geothermal energy an ideal complement to other renewable sources, helping to balance the grid and ensure a reliable power supply. By integrating geothermal energy into the grid, countries can reduce their reliance on fossil fuels and enhance the resilience of their energy systems.

The integration of geothermal energy into national grids requires robust infrastructure capable of accommodating the unique characteristics of geothermal power. Grid operators must ensure that transmission and distribution networks are equipped to handle the steady output of geothermal plants, as well as the potential for increased capacity as new projects come online. Upgrading and modernizing grid infrastructure is essential to facilitate the seamless integration of geothermal energy and maximize its contribution to the energy mix.

Regulatory frameworks play a crucial role in supporting the integration of geothermal energy into national grids. Governments must establish clear policies and incentives that encourage the development of geothermal projects and facilitate their connection to the grid. This includes streamlining permitting processes, providing financial incentives for renewable energy development, and setting targets for geothermal capacity. By creating a supportive regulatory environment, governments can attract investment in geothermal energy and accelerate its deployment.

Technological advancements are also key to the successful integration of geothermal energy into national grids. Innovations in grid management and control systems enable operators to optimize the flow of electricity from geothermal plants, ensuring that power is delivered efficiently and reliably. Advanced forecasting and monitoring tools can help grid operators anticipate changes in demand and supply, allowing for more effective management of geothermal resources. By leveraging technology, countries can enhance the flexibility and resilience of their energy systems, making it easier to integrate geothermal energy and other renewables.

The integration of geothermal energy into national grids also presents opportunities for economic development and job creation. The construction and operation of geothermal power plants require skilled labor and specialized expertise, creating employment opportunities in engineering, construction, and maintenance. Additionally, the development of geothermal resources can stimulate local economies by attracting investment and supporting infrastructure development. By investing in geothermal energy, countries can drive economic growth while advancing their renewable energy goals.

Community engagement and public acceptance are important factors in the successful integration of geothermal energy into national grids. Building public support for geothermal projects requires transparent communication about the benefits and potential impacts of geothermal energy. Engaging with local communities and stakeholders throughout the project development process can foster trust and collaboration, ensuring that projects align with local needs and priorities. Public education and awareness campaigns can also enhance the understanding of geothermal energy and its role in a sustainable energy future.

International collaboration and knowledge exchange are essential for advancing the integration of geothermal energy into national grids. Countries can learn from the experiences of others that have successfully integrated geothermal energy, sharing best practices and lessons learned. International organizations and partnerships can facilitate the exchange of technical expertise and provide support for capacity-building initiatives. By working together, countries can overcome common challenges and accelerate the global transition to renewable energy.

The integration of geothermal energy into national grids is a complex and multifaceted process, but the potential benefits are significant. By harnessing the Earth's natural heat, countries can enhance energy security, reduce greenhouse gas emissions, and support economic development. As the world continues to seek solutions to the challenges of climate change and energy transition, geothermal energy offers a reliable and sustainable option that can play a vital role in achieving a cleaner and more resilient energy future. The journey towards integrating geothermal energy into national grids is one of innovation and collaboration, driven by the shared goal of creating a sustainable energy system for generations to come.

As countries advance in their efforts to integrate geothermal energy into national grids, the importance of strategic planning and coordination cannot be overstated. Effective integration requires a comprehensive approach that considers the entire energy ecosystem, from resource assessment and project development to grid management and policy implementation. By adopting a holistic perspective, countries can ensure that geothermal energy is seamlessly incorporated into their energy systems, maximizing its benefits and minimizing potential challenges.

One critical aspect of strategic planning is the identification and assessment of geothermal resources. Conducting thorough geological surveys and feasibility studies is essential to determine the viability of geothermal projects and their potential contribution to the grid. By understanding the characteristics and capacity of geothermal resources, countries can prioritize projects that offer the greatest potential for energy generation and grid stability. This targeted approach allows for the efficient allocation of resources and investment, ensuring that geothermal energy is developed in a way that aligns with national energy goals.

Collaboration between government agencies, industry stakeholders, and research institutions is vital for the successful integration of geothermal energy. By fostering partnerships and encouraging dialogue, countries can leverage the expertise and resources of various stakeholders to drive innovation and development. Collaborative efforts can lead to the creation of standardized practices and guidelines for geothermal integration, facilitating the sharing of knowledge and best practices across the industry. This collective approach enhances the efficiency and effectiveness of geothermal projects, ensuring that they contribute positively to the national grid.

The role of policy and regulation in geothermal integration cannot be underestimated. Governments must establish clear and consistent policies that support the development and integration of geothermal energy. This includes setting ambitious targets for renewable energy adoption, providing financial incentives for geothermal projects, and implementing regulatory frameworks that facilitate grid connection. By creating a stable and predictable policy environment, governments can attract investment and encourage the growth of the geothermal sector, ensuring that it plays a significant role in the national energy mix.

www.ingramcontent.com/pod-product-compliance
Lightning Source LLC
Chambersburg PA
CBHW072034150726
47999CB00002B/897